Reteach Book

Grade 1

PROVIDES   Tier 1 Intervention for Every Lesson

 HOUGHTON MIFFLIN HARCOURT

Contents

CRITICAL AREA: Operations and Algebraic Thinking

Chapter 1: Addition Concepts

Chapter 2: Subtraction Concepts

Chapter 3: Addition Strategies

Chapter 4: Subtraction Strategies

Chapter 5: Addition and Subtraction Relationships

CRITICAL AREA: Number and Operations in Base Ten

CRITICAL AREA: Measurement and Data

Chapter 9: Measurement

Chapter 10: Represent Data

CRITICAL AREA: Geometry

Chapter 11: Three-Dimensional Geometry

Chapter 12: Two-Dimensional Geometry

Algebra • Use Pictures to Add To

COMMON CORE STANDARD CC.1.OA.1
Represent and solve problems involving addition and subtraction.

3 cows   and   2 more cows   __5__ cows.

Draw circles around the animals added to the group. Write how many.

1.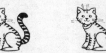

3 cats   and   4 more cats   ____ cats

2.

2 bees   and   2 more bees   ____ bees

3.

5 dogs   and   1 more dog   ____ dogs

Model Adding To

COMMON CORE STANDARD CC.1.OA.1
Represent and solve problems involving addition and subtraction.

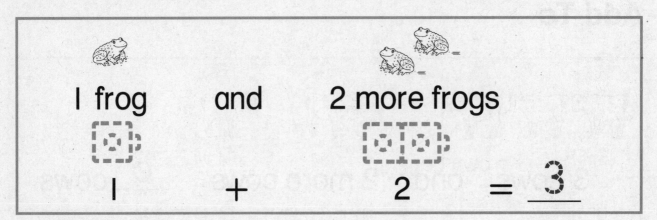

I frog   and   2 more frogs

1   +   2   = 3

Use ⬚ **to show adding to. Draw the** ⬚.
Write the sum.

1. 3 horses and 4 more horses

3   +   4   = ___

2. I bee and I more bee

I   +   I   = ___

3. 4 cows and I more cow

4   +   I   = ___

Model Putting Together

COMMON CORE STANDARD CC.1.OA.1
Represent and solve problems involving
addition and subtraction.

Use  to add two groups.
Put the groups together to
find how many.

There are 3 brown dogs.

There is 1 white dog.

How many dogs are there?

 dogs

$3 + 1 = 4$

Use  to solve. Draw to show your work.
Write how many.

1. There are 4 black bears and
   3 brown bears. How many
   bears are there?

   _____ bears

   $4 + 3 =$ _____

2. There are 6 red flowers and
   2 white flowers. How many
   flowers are there?

   _____ flowers

   $6 + 2 =$ _____

Problem Solving • Model Addition

COMMON CORE STANDARD CC.1. OA.1
Represent and solve problems involving addition and subtraction.

Rico has 3 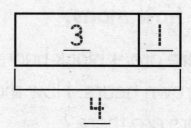. Then he gets 1 more 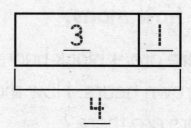.
How many 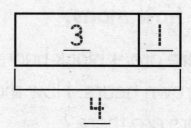 does he have now?

Unlock the Problem

What do I need to find?	**What information do I need to use?**
the number of **crayons** Rico has now.	Rico has ___**3**___ 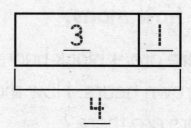. He gets ___**1**___ 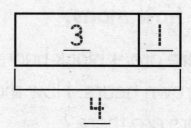.

Show how to solve the problem.

3	1

4

$3 + 1 =$ ___

Read the problem. Use the bar model to solve.
Complete the model and the number sentence.

1. There are 5 birds flying.
   Then 3 more birds join them.
   How many birds are flying now?

5	3

___

$5 + 3 =$ ___

Algebra • Add Zero

COMMON CORE STANDARD CC.1.OA.3
Understand and apply properties of
operations and the relationship between
addition and subtraction.

Use ⬤ to show each number.
Add. Write the sum.

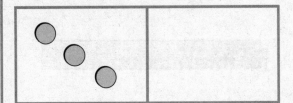 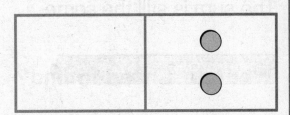

$3 + 0 = $ _3_ $0 + 2 = $ _2_

When you add zero to a number,
the sum is that number.

Use ⬤ to show each number.
Draw the ⬤. Write the sum.

1.

$0 + 4 = $ ___

2.

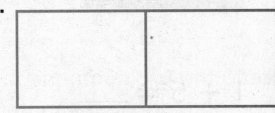

$6 + 0 = $ ___

3.

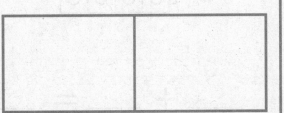

$0 + 1 = $ ___

4.

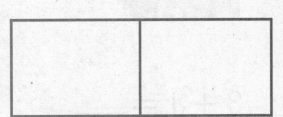

$0 + 5 = $ ___

Algebra · Add in Any Order

COMMON CORE STANDARD CC.1.OA.3
Understand and apply properties of
operations and the relationship between
addition and subtraction.

Write an addition sentence.
Change the order of the addends.
The sum is still the same.

Turn the cube
train around.

$5 + 3 = \underline{}8$
$\underline{\text{sum}}$

$\underline{}3 + \underline{}5 = \underline{}8$
$\underline{\text{sum}}$

Use 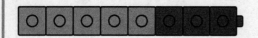 to add. Write
the sum.

Change the order of the
addends. Color to match.
Write the addition sentence.

1.

$1 + 5 = \underline{\qquad}$

$\underline{\qquad} + \underline{\qquad} = \underline{\qquad}$

2.

$3 + 1 = \underline{\qquad}$

$\underline{\qquad} + \underline{\qquad} = \underline{\qquad}$

Algebra · Put Together Numbers to 10

COMMON CORE STANDARD CC.1.OA.1
Represent and solve problems involving
addition and subtraction.

You can use  to model ways to make 7.

$6 + \underline{1} = 7$

$5 + \underline{2} = 7$

Use . Draw to show how to make 7.
Complete the addition sentences.

1. ◯◯◯◯

$4 + \underline{} = 7$

2. ◯◯◯

$3 + \underline{} = 7$

3. ◯◯

$2 + \underline{} = 7$

4. ◯

$1 + \underline{} = 7$

Addition to 10

COMMON CORE STANDARD CC.1.OA.6
Add and subtract within 20.

You can use ⬜ to help you add.

$$4 \quad \square\square\square\square$$
$$+ 2 \quad \square\square$$
$$\overline{6}$$

$$6 \quad \square\square\square\square\square\square$$
$$+ 3 \quad \square\square\square$$

Use ⬜. Write the sum.

1.
$$1 \quad \square$$
$$+ 2 \quad \square\square$$

2.
$$4 \quad \square\square\square\square$$
$$+ 1 \quad \square$$

3.
$$3 \quad \square\square\square$$
$$+ 5 \quad \square\square\square\square\square$$

4.
$$2 \quad \square\square$$
$$+ 3 \quad \square\square\square$$

Use Pictures to Show Taking From

COMMON CORE STANDARD CC.1.OA.1
Represent and solve problems involving addition and subtraction.

Use the picture.

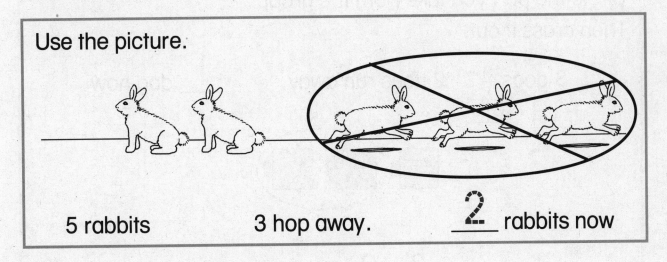

5 rabbits   3 hop away.   __2__ rabbits now

Write how many there are now.

1.

8 birds   4 fly away.   _____ birds now

2.

7 bees   2 fly away.   _____ bees now

Model Taking From

COMMON CORE STANDARD CC.1.OA.1
Represent and solve problems involving
addition and subtraction.

Circle the part you take from the group.
Then cross it out.

3 dogs   2 dogs run away. |‾ dog now

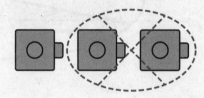

$3 - 2 = $ |‾

**Circle the part you take from the group.
Then cross it out. Write the difference.**

1. 4 goats   2 goats walk away.   ____ goats now

$4 - 2 = $ ___

2. 6 ants   3 ants walk away.   ____ ants now

$6 - 3 = $ ___

Model Taking Apart

COMMON CORE STANDARD CC.1.OA.1
Represent and solve problems involving
addition and subtraction.

You can use ◯ to **subtract**.
Sam has 6 cars. 4 cars are red.
The rest are yellow.
How many cars are yellow?

_2_ cars are yellow.

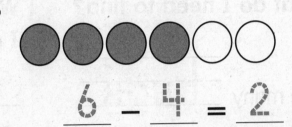

$$6 - 4 = 2$$

**Use ◯ to solve. Color. Write the number sentence
and how many.**

1. There are 5 books.
   1 book is red. The
   rest are yellow. How
   many books are yellow?

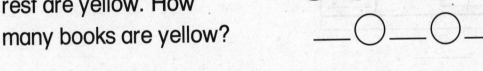

____ yellow books

2. There are 6 blocks.
   3 blocks are small.
   The rest are big.
   How many blocks
   are big?

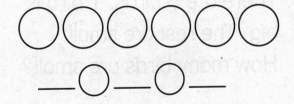

____ big blocks

Problem Solving • Model Subtraction

COMMON CORE STANDARD CC.1.OA.1
Represent and solve problems involving
addition and subtraction.

There were 9 bugs on a rock. 7 bugs ran away.
How many bugs are on the rock now?

What do I need to find?	What information do I need to use?
how many __bugs__ on the rock now	_9_ bugs on a rock _7_ bugs ran away

Show how to solve the problem.

7	2

9

$9 - 7 = \underline{2}$

Read the problem. Use the model to solve.
Complete the model and the number sentence.

1. There are 5 birds. 1 bird is
   big. The rest are small.
   How many birds are small?

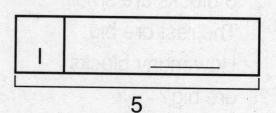

1	

5

$5 - 1 = \underline{}$

Use Pictures and Subtraction to Compare

COMMON CORE STANDARD CC.1.OA.8
Work with addition and subtraction equations.

You can subtract to compare groups.

$$7 - 6 = \underline{}$$

There is I **more** than there are 🏀.

There is I **fewer** 🏀 than there are .

Subtract to compare.

1.

$$5 - 3 = \underline{}$$

____ more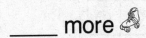

2.

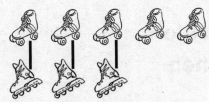

$$6 - 4 = \underline{}$$

____ fewer 🪁

3.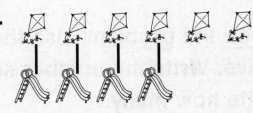

$$4 - 1 = \underline{}$$

____ more ⬭

4.

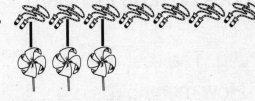

$$7 - 3 = \underline{}$$

____ fewer 🌀

Subtract to Compare

COMMON CORE STANDARD CC.1.OA.1
Represent and solve problems involving
addition and subtraction.

You can use to show the bar model.

8

6

Andy has 8 balloons.
Jill has 6 balloons.
How many more balloons
does Andy have than Jill?

8

6

2

_____ more balloons

8 – 6 = 2

Read the problem. Use the bar model to solve. Write the number sentence. Then write how many.

1. Bo has 6 rocks.
   Jen has 4 rocks.
   How many more rocks
   does Bo have than Jen?

6

4

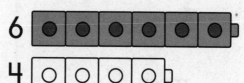

6

4

_____

_____ more rocks

____ – ____ = ____

Subtract All or Zero

COMMON CORE STANDARD CC.1.OA.8
Work with addition and subtraction equations

When you subtract zero from a number, the difference is the number.

No ⬤ are crossed out.

$4 - 0 = \underline{4}$

When you subtract a number from itself, the difference is zero.

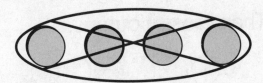

All ⬤ are crossed out.

$4 - 4 = \underline{0}$

Use ⬤. Write the difference.

1.

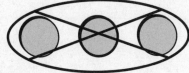

$3 - 3 = \underline{0}$

2.

$5 - 0 = \underline{}$

3.

$2 - 0 = \underline{}$

4.

$1 - 1 = \underline{}$

5.

$6 - 0 = \underline{}$

6.

$4 - 4 = \underline{}$

Algebra • Take Apart Numbers

COMMON CORE STANDARD CC.1.OA.1
Represent and solve problems involving addition and subtraction.

You can use ⬤ to take apart 6.

Circle the part you take away.
Then cross it out.

$6 - 5 = \underline{1}$

$6 - 4 = \underline{2}$

Use ⬤ to take apart 6. Circle the part you take away. Then cross it out. Complete the subtraction sentence.

1. ○○○○○○ $6 - 3 = \underline{}$

2. ○○○○○○ $6 - 2 = \underline{}$

3. ○○○○○○ $6 - 1 = \underline{}$

4. ○○○○○○ $6 - 0 = \underline{}$

Subtraction from 10 or Less

COMMON CORE STANDARD CC.1.OA.6
Add and subtract within 20.

You can use  to help you subtract.

$$\begin{array}{r} 6 \\ -3 \\ \hline \boxed{3} \end{array}$$

$$\begin{array}{r} 3 \\ -1 \\ \hline \boxed{2} \end{array}$$

Write the subtraction problem.

1.
$$\begin{array}{r} 7 \\ -4 \\ \hline \boxed{} \end{array}$$

2.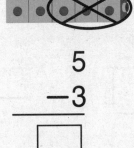
$$\begin{array}{r} 5 \\ -3 \\ \hline \boxed{} \end{array}$$

3.
$$\begin{array}{r} 8 \\ -1 \\ \hline \boxed{} \end{array}$$

4.
$$\begin{array}{r} 4 \\ -2 \\ \hline \boxed{} \end{array}$$

Algebra • Add in Any Order

COMMON CORE STANDARD CC.1.OA.3
Understand and apply properties of operations and the relationship between addition and subtraction.

You can change the order of the addends.
The sum is the same.

$$\begin{array}{r} 5 \\ + 2 \\ \hline 7 \end{array}$$

$$\begin{array}{r} 2 \\ + 5 \\ \hline 7 \end{array}$$

Add. Change the order of the addends. Add again.

1.

$$\begin{array}{r} 3 \\ + 1 \\ \hline \end{array} \qquad \begin{array}{r} \\ + \\ \hline \end{array}$$

2.

$$\begin{array}{r} 4 \\ + 2 \\ \hline \end{array} \qquad \begin{array}{r} \\ + \\ \hline \end{array}$$

3.

$$\begin{array}{r} 8 \\ + 3 \\ \hline \end{array} \qquad \begin{array}{r} \\ + \\ \hline \end{array}$$

4.

$$\begin{array}{r} 9 \\ + 5 \\ \hline \end{array} \qquad \begin{array}{r} \\ + \\ \hline \end{array}$$

Count On

You can count on to find $4 + 3$.
Start with the greater addend.
Then count on. Write the sum.

To add 3,
count on 3.

○ ○ ○

| 4 | 5 | 6 | 7 |

$4 + 3 = \underline{7}$

Circle the greater addend. Count on 1, 2, or 3. Write the missing numbers.

1. $1 + 6$

○ ○

| 6 | ___ |

$1 + 6 = \underline{}$

2. $9 + 1$

○ ○

| 9 | ___ |

$9 + 1 = \underline{}$

3. $4 + 2$

○ ○ ○

| 4 | ___ ___ |

$4 + 2 = \underline{}$

4. $3 + 8$

○ ○ ○ ○

| 8 | ___ ___ ___ |

$3 + 8 = \underline{}$

Add Doubles

COMMON CORE STANDARD CC.1.OA.6
Add and subtract within 20.

The addends are the same in a doubles fact.

$\underline{3} + \underline{3} = 6$

Draw **to show the addends.**
Write the missing numbers.

1.

___ + ___ = 8

2.

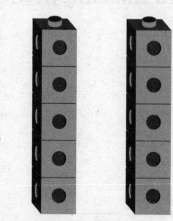

___ + ___ = 10

3.

___ + ___ = 4

4.

___ + ___ = 2

COMMON CORE STANDARD CC.1.OA.6
Add and subtract within 20.

Use Doubles to Add

Use a doubles fact to solve 4 + 3.
Break apart 4 into 1 + 3.

○ ● ● ● ● ● ●

1   +   3   +   3

1   +   6   = 7

So, 4 + 3 = __7__.

> **THINK**
> 3 + 3 = 6.
> I more than 6 is 7.

Use ○ ● to model. Break apart
to make a doubles fact. Add.

1. 6 + 5

○ ● ● ● ● ● ● ● ● ● ●

__1__ + ___ + ___

___ + ___ = ___

So, 6 + 5 = ___.

2. 8 + 7

○ ● ● ● ● ● ● ● ● ● ● ● ● ● ● ●

___ + ___ + ___

___ + ___ = ___

So, 8 + 7 = ___.

Doubles Plus 1 and Doubles Minus 1

COMMON CORE STANDARD CC.1.OA.6
Add and subtract within 20.

You can use doubles plus one facts and doubles minus one to add.

Use doubles fact 3 + 3 = 6.

doubles plus one

doubles minus one

$$3 + 4 = \underline{7}$$

$$3 + 2 = \underline{5}$$

Use doubles plus one or doubles minus one to add.

1.

$$5 + 6 = \underline{}$$

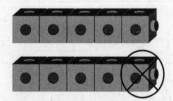

$$5 + 4 = \underline{}$$

2.

$$2 + 3 = \underline{}$$

$$2 + 1 = \underline{}$$

Add 10 and More

COMMON CORE STANDARD CC.1.OA.6
Add and subtract within 20.

You can use counters and a ten frame to add
a number to 10.

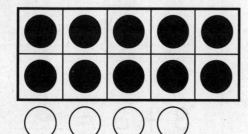

Find 10 + 4.

$$\begin{array}{r} 10 \\ + 4 \\ \hline 14 \end{array}$$

**Draw ◯. Show the number that is added to
10. Write the sum.**

1.

$$\begin{array}{r} 10 \\ + 3 \\ \hline \end{array}$$

2.

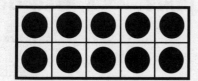

$$\begin{array}{r} 10 \\ + 7 \\ \hline \end{array}$$

Practice the Strategies

COMMON CORE STANDARD CC.1.OA.6
Add and subtract within 20.

You can use different addition strategies to find sums.

Count On

$6 + 2 = \underline{8}$

Doubles

$3 + 3 = \underline{6}$

Doubles Plus 1

$5 + 6 = \underline{11}$

Doubles Minus 1

$5 + 4 = \underline{9}$

---

1. Count on 1.	2. Count on 2.	3. Count on 3.
$7 + 1 = \underline{\hphantom{0}}$	$7 + 2 = \underline{\hphantom{0}}$	$7 + 3 = \underline{\hphantom{0}}$
4. Use doubles.	5. Use doubles plus 1.	6. Use doubles minus 1.
$6 + 6 = \underline{\hphantom{0}}$	$6 + 7 = \underline{\hphantom{0}}$	$6 + 5 = \underline{\hphantom{0}}$

Make a 10 to Add

COMMON CORE STANDARD CC.1.OA.6
Add and subtract within 20.

Show 8 + 5 with counters and a ten frame.

Use ◯.

| 8 |
| 5 |

Make a ten. Add.

10
+ 3
13

So, 8 + 5 = 13.

Draw ◯ to show the second addend.
Make a ten. Add.

1. 8 + 6

| 8 |
| 6 |

10
+ 4

So, 8 + 6 = ___.

2. 9 + 7

| 9 |
| 7 |

10
+ 6

So, 9 + 7 = ___.

Use Make a 10 to Add

What is 9 + 5? Make a 10 to add.

Use ◯ and a ten frame.
Show the addends.

| 9 |
| 5 |

Make a 10.
Add.

| 10 |
| 4 |

So, 9 + 5 = __14__.

Show the greater addend in the ten frame.

Draw ◯. Make a ten to add.

1. 8 + 5

| 8 |
| 5 |

| 10 |
| 3 |

10
+ 3

So, 8 + 5 = ____.

2. 7 + 4

| 7 |
| 4 |

| 10 |
| 1 |

10
+ 1

So, 7 + 4 = ____.

Algebra • Add 3 Numbers

COMMON CORE STANDARD CC.1.OA.3
Understand and apply properties of operations and the relationship between addition and subtraction.

You can add numbers in any order.

$3 + 4 + 1 = 8$

$(3) + (4) + 1 = 8$ $\qquad$ $3 + (4) + (1) = 8$

$\underline{7} + 1 = 8$ $\qquad$ $3 + \underline{5} = 8$

Use circles to change which two addends you add first. Complete the addition sentences.

1. $(2) + (1) + 8 = 11$ $\qquad$ $2 + (1) + (8) = 11$

$\underline{} + 8 = 11$ $\qquad$ $2 + \underline{} = 11$

2. $(7) + (2) + 3 = 12$ $\qquad$ $7 + (2) + (3) = 12$

$\underline{} + 3 = 12$ $\qquad$ $7 + \underline{} = 12$

Algebra • Add 3 Numbers

COMMON CORE STANDARD CC.1.OA.3
Understand and apply properties of operations and the relationship between addition and subtraction.

What strategies help you add 3 numbers?

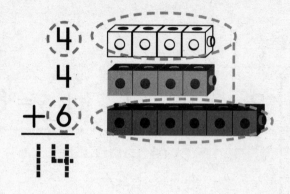

$$\begin{array}{r} 4 \\ 4 \\ +6 \\ \hline 14 \end{array}$$

$$\begin{array}{r} 4 \\ 4 \\ +6 \\ \hline 14 \end{array}$$

4 + 6 make a 10.

$\boxed{10} + 4 = \boxed{14}$

4 + 4 = 8 is a doubles fact.

$\boxed{8} + 6 = \boxed{14}$

**Choose a strategy. Circle two addends
to add first. Write the sum.
Then find the total sum.**

1.

$$\begin{array}{r} 7 \\ 3 \\ +3 \\ \hline 13 \end{array}$$
$\boxed{10}$

2.
$$\begin{array}{r} 2 \\ 2 \\ +8 \\ \hline \end{array}$$
$\boxed{}$

3.
$$\begin{array}{r} 4 \\ 3 \\ +3 \\ \hline \end{array}$$
$\boxed{}$

4.
$$\begin{array}{r} 5 \\ 5 \\ +4 \\ \hline \end{array}$$
$\boxed{}$

Problem Solving • Use Addition Strategies

COMMON CORE STANDARD CC.1.OA.2
Represent and solve problems involving addition and subtraction.

Tory has 9 toys. Bob has 4 toys.
Joy has 2 toys. How many toys
do they have?

Unlock the Problem

What do I need to find?	**What information do I need to use?**
how many **toys** they have	Tory has **9** toys. Bob has **4** toys. Joy has **2** toys.

Show how to solve the problem.

____ (+) ____ (+) ____ (=) ____

_____ toys

Draw a picture to solve.

1. Rick has 7 books.
   He gets 2 more books.
   He then gets 2 more books.
   How many books does
   Rick have now?

_____ books

Name _____

Count Back

COMMON CORE STANDARD CC.1.OA.5
Add and subtract within 20.

Count back to subtract.

Use 9 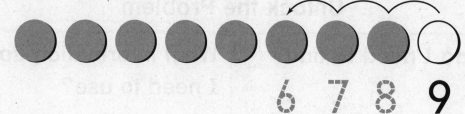. Count back 3.

This shows counting back 3 from 9.

$$\underset{6}{\text{\_\_}} \quad \underset{7}{\text{\_\_}} \quad \underset{8}{\text{\_\_}} \quad 9$$

$9 - 3 = \underline{6}$

Use 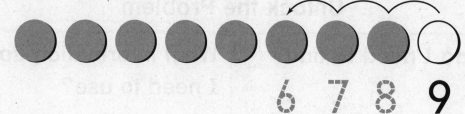. Count back 1, 2, or 3.

Write the difference.

1. $5 - 1 = \underline{}$

$$\underset{}{\text{\_\_}} \quad 5$$

2. $7 - 2 = \underline{}$

$$\underset{}{\text{\_\_}} \quad \underset{}{\text{\_\_}} \quad 7$$

3. $6 - 3 = \underline{}$

$$\underset{}{\text{\_\_}} \quad \underset{}{\text{\_\_}} \quad \underset{}{\text{\_\_}} \quad 6$$

Think Addition to Subtract

COMMON CORE STANDARD CC.1.OA.4
Understand and apply properties of operations and the relationship between addition and subtraction.

What is $7 - 4$?

Think $4 + \underline{\ 3\ } = 7$

So $7 - 4 = \underline{\ 3\ }$

Use  to model the number sentences.
Draw 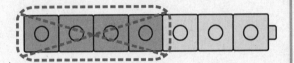 to show your work.

1. What is $11 - 2$?

Think $2 + \underline{\ \ \ } = 11$

So $11 - 2 = \underline{\ \ \ }$

2. What is $10 - 6$?

Think $6 + \underline{\ \ \ } = 10$

So $10 - 6 = \underline{\ \ \ }$

3. What is $6 - 1$?

Think $1 + \underline{\ \ \ } = 6$

So $6 - 1 = \underline{\ \ \ }$

Use Think Addition to Subtract

COMMON CORE STANDARD CC.1.OA.4
Understand and apply properties of
operations and the relationship between
addition and subtraction.

Think of an addition fact to help you subtract.	Think

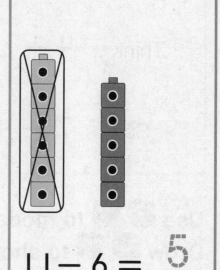

$11 - 6 = \underline{\ ?\ }$ | $6 + \underline{\ 5\ } = 11$ | $11 - 6 = \underline{\ 5\ }$

Use an addition fact to help you subtract.

1. What is $9 - 4$?

Use $\quad 4 + \underline{\quad} = 9$

So $\quad 9 - 4 = \underline{\quad}$

2. What is $10 - 6$?

Use $\quad 6 + \underline{\quad} = 10$

So $\quad 10 - 6 = \underline{\quad}$

3. What is $12 - 5$?

Use $\quad 5 + \underline{\quad} = 12$

So $\quad 12 - 5 = \underline{\quad}$

4. What is $8 - 5$?

Use $\quad 5 + \underline{\quad} = 8$

So $\quad 8 - 5 = \underline{\quad}$

COMMON CORE STANDARD CC.1.OA.6
Add and subtract within 20.

Use 10 to Subtract

Find 14 − 9.

Start with 9 cubes.

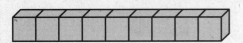

Make a 10.

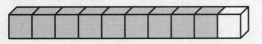

Add cubes to make 14.

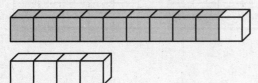

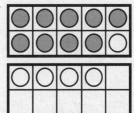

Count what you added.

You added ___5___.

So, 14 − 9 = ___5___.

Use ▣. Make a ten to subtract.
Draw to show your work.

1. 12 − 8 = ___?___

2. 15 − 9 = ___?___

   12 − 8 = ___

   15 − 9 = ___

Break Apart to Subtract

COMMON CORE STANDARD CC.1.OA.6
Add and subtract within 20.

What is 14 − 5?

Start with 14. Make a ten.

Take __4__ from 14.

$14 - 4 = 10$

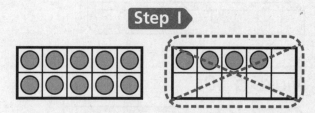

Step 1

Then take __1__ more.

$10 - 1 = 9$

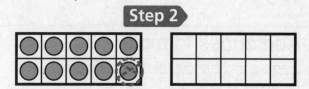

Step 2

So, 14 − 5 = __9__

Subtract.

1. What is 17 − 9?

Take 7 counters from 17.

$17 - 7 = $ ____

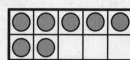

Step 1

Then take ____ counters from 10.

____ − ____ = ____

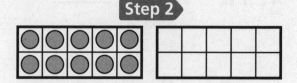

Step 2

So, 17 − 9 = ____

Problem Solving · Use Subtraction Strategies

COMMON CORE STANDARD CC.1.OA.1
Represent and solve problems involving addition and subtraction.

Lara has 15 crackers. She gives
some of them away. She has 8 left.
How many crackers does she give away?

Unlock the Problem

What do I need to find?	**What information do I need to use?**
how many crackers	Lara has 15 crackers.
Lara gives away	Lara has 8 crackers left.

Show how to solve the problem.

◯ ◯ ◯ ◯ ◯ ◯ ◯ ◯ ◯ ◯ ◯ ◯ ◯ ◯ ◯

Lara gives away __7__ crackers.

Act it out to solve. Draw to show your work.

1. Min has 13 marbles.
   She gives some away.
   She has 5 left.
   How many marbles does
   she give away?

Min gives away _____ marbles.

COMMON CORE STANDARD CC.1.OA.1
Represent and solve problems involving
addition and subtraction.

Problem Solving • Add or Subtract

There are 12 skunks in the woods.

Some skunks walk away.

There are 8 skunks still in the woods.

How many skunks walk away?

Unlock the Problem

What do I need to find?	**What information do I need to use?**
how many walk away _ _ _ **skunks** _ _ _	**12** skunks in the woods **8** skunks still in the woods

Show how to solve the problem.

	8

12

4 walk away

8 skunks still in the woods

12 skunks

Make a model to solve.
Use 🎲🎲 to help you.

I. There are 15 frogs on a log.

Some frogs hop away.

There are 7 frogs still on the log.

How many frogs hop away?

_____ frogs hop away

7	

15

Record Related Facts

COMMON CORE STANDARD CC.1.OA.6
Add and subtract within 20.

Use the numbers to
write four related facts.

$$6 + 4 = 10 \qquad 10 - 4 = 6$$

THINK
Each number is
in all four facts.

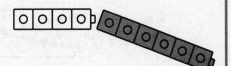

$$4 + 6 = 10 \qquad 10 - 6 = 4$$

Use the numbers to make related facts.

1.

$$6 + 8 = 14 \qquad \boxed{} - 8 = 6$$
$$8 + \boxed{} = 14 \qquad 14 - 6 = \boxed{}$$

2.

$$\boxed{} + 7 = 9 \qquad 9 - \boxed{} = 2$$
$$7 + 2 = 9 \qquad \boxed{} - 2 = 7$$

3.

$$5 + \boxed{} = 11 \qquad \boxed{} - 6 = 5$$
$$6 + 5 = 11 \qquad 11 - \boxed{} = 6$$

4.

$$3 + 9 = 12 \qquad \boxed{} - 9 = 3$$
$$\boxed{} + 3 = 12 \qquad 12 - \boxed{} = 9$$

Identify Related Facts

COMMON CORE STANDARD CC.1.OA.6
Add and subtract within 20.

If you know an addition fact, you will also know the related subtraction fact.

> Both facts use 2, 4, and 6. They are related facts.

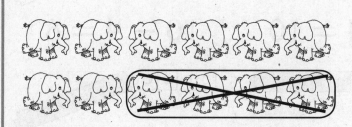

$$2 \oplus 4 \ominus 6$$

$$6 \ominus 4 \ominus 2$$

Add and subtract the related facts.

1.

$$7 + 8 = \underline{}$$

$$15 - 8 = \underline{}$$

2.

$$7 + 4 = \underline{}$$

$$11 - 4 = \underline{}$$

3.

$$1 + 8 = \underline{}$$

$$9 - 8 = \underline{}$$

Name _____

COMMON CORE STANDARD CC.1.OA.6
Add and subtract within 20.

Use Addition to Check Subtraction

You can use addition to check subtraction.

You start with 8.
Take apart to subtract.

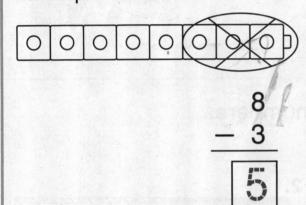

$$\begin{array}{r} 8 \\ -\ 3 \\ \hline 5 \end{array}$$

THINK
Put the 5 and
3 back together.

Add to check.
You end with 8.

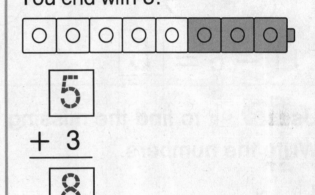

$$\begin{array}{r} 5 \\ +\ 3 \\ \hline 8 \end{array}$$

Use 🎲 🎲 to help you. Subtract.
Then add to check your answer.

$$\begin{array}{r} 7 \\ -\ 3 \\ \hline \square \end{array}$$

$$\begin{array}{r} \square \\ +\ 3 \\ \hline \square \end{array}$$

Algebra • Missing Numbers

COMMON CORE STANDARD CC.1.OA.8
Work with addition and subtraction equations.

Add or subtract to find the missing numbers.

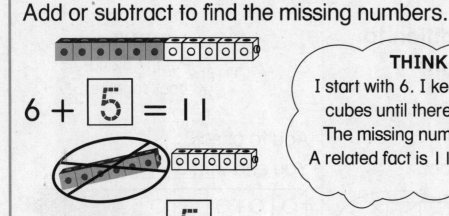

$$6 + \boxed{5} = 11$$

THINK
I start with 6. I keep adding cubes until there are 11. The missing number is 5. A related fact is $11 - 6 = 5$.

$$11 - 6 = \boxed{5}$$

Use 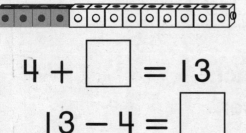 to find the missing numbers.

Write the numbers.

1.

$$4 + \boxed{} = 13$$

$$13 - 4 = \boxed{}$$

2.

$$7 + \boxed{} = 15$$

$$15 - 7 = \boxed{}$$

3.

$$8 + \boxed{} = 14$$

$$14 - 8 = \boxed{}$$

4.

$$9 + \boxed{} = 16$$

$$16 - 9 = \boxed{}$$

5.

$$9 + \boxed{} = 18$$

$$18 - 9 = \boxed{}$$

6.

$$8 + \boxed{} = 16$$

$$16 - 8 = \boxed{}$$

Record Related Facts

COMMON CORE STANDARD CC.1.OA.6
Add and subtract within 20.

Use the numbers to
write four related facts.

$$6 + 4 = 10$$

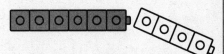

$$10 - 4 = 6$$

THINK
Each number is
in all four facts.

$$4 + 6 = 10$$

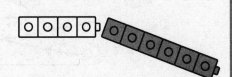

$$10 - 6 = 4$$

Use the numbers to make related facts.

1.   6 8 14

$$6 + 8 = 14 \qquad \boxed{} - 8 = 6$$

$$8 + \boxed{} = 14 \qquad 14 - 6 = \boxed{}$$

2.   2 7 9

$$\boxed{} + 7 = 9 \qquad 9 - \boxed{} = 2$$

$$7 + 2 = 9 \qquad \boxed{} - 2 = 7$$

3.   5 6 11

$$5 + \boxed{} = 11 \qquad \boxed{} - 6 = 5$$

$$6 + 5 = 11 \qquad 11 - \boxed{} = 6$$

4.   3 9 12

$$3 + 9 = 12 \qquad \boxed{} - 9 = 3$$

$$\boxed{} + 3 = 12 \qquad 12 - \boxed{} = 9$$

COMMON CORE STANDARD CC.1.OA.6
Add and subtract within 20.

Identify Related Facts

If you know an addition fact, you will also know the related subtraction fact.

> Both facts use 2, 4, and 6. They are related facts.

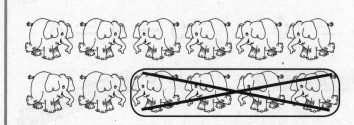

$2 \oplus 4 = 6$

$6 \ominus 4 = 2$

Add and subtract the related facts.

1.

$7 + 8 = \underline{\hphantom{00}}$

$15 - 8 = \underline{\hphantom{00}}$

2.

$7 + 4 = \underline{\hphantom{00}}$

$11 - 4 = \underline{\hphantom{00}}$

3.

$1 + 8 = \underline{\hphantom{00}}$

$9 - 8 = \underline{\hphantom{00}}$

Algebra • Use Related Facts

COMMON CORE STANDARD CC.1.OA.8
Work with addition and subtraction equations.

Find $11 - 6$.

Use counters to help you.

THINK
Start with 6. How many do I add to make 11?

$6 + \underline{5} = 11$

$11 - 6 = \underline{5}$

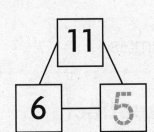

Use counters. Write the missing numbers.

1. Find $13 - 8$.

$8 + \underline{} = 13$

$13 - 8 = \underline{}$

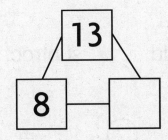

2. Find $12 - 3$.

$3 + \underline{} = 12$

$12 - 3 = \underline{}$

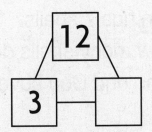

Choose an Operation

COMMON CORE STANDARD CC.1.OA.1
Represent and solve problems involving
addition and subtraction.

Liz has 15 stuffed animals.
She gives away 6. How many
stuffed animals are left?

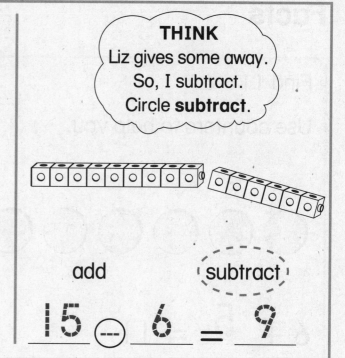

THINK
Liz gives some away.
So, I subtract.
Circle **subtract**.

add   (subtract)

_9_ stuffed animals

_15_ (−) _6_ = _9_

Circle add or subtract.
Write a number sentence to solve.

1. Misha has 11 crackers.
   He eats 2 crackers.
   How many crackers
   are left?

   add   subtract

   _____ crackers   _____ ◯ _____ = _____

2. Lynn has 5 shells.
   Dan has 7 shells.
   How many shells do
   Lynn and Dan have?

   add   subtract

   _____ shells   _____ ◯ _____ = _____

Algebra · Equal and Not Equal

COMMON CORE STANDARD CC.1.OA.7
Work with addition and subtraction equations.

An equal sign means both sides are the same.

$$3 + 3 = 6 - 0$$

THINK
$3 + 3 = 6$ and $6 - 0 = 6$.
Is 6 the same as 6?

yes

It is true.

$$3 + 2 = 5 - 2$$

THINK
$3 + 2 = 5$ and $5 - 2 = 3$.
Is 5 the same as 3?

_no_

It is false.

Which is true? Circle your answer.
Which is false? Cross out your answer.

1. $7 - 5 = 5 - 2$

   $8 - 8 = 6 - 6$

2. $1 + 8 = 18$

   $2 + 8 = 8 + 2$

3. $4 + 3 = 5 + 2$

   $7 + 3 = 4 + 5$

4. $9 - 2 = 9 + 2$

   $9 = 10 - 1$

Algebra • Ways to Make Numbers to 20

COMMON CORE STANDARD CC.1.OA.6
Add and subtract within 20.

These are some ways to make the number 14.

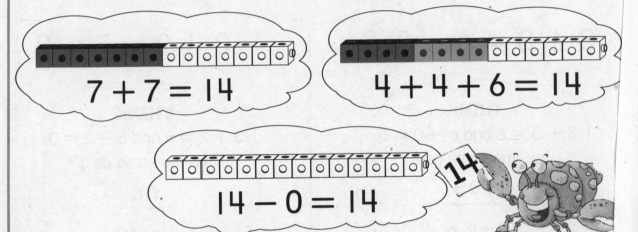

$$7 + 7 = 14$$

$$4 + 4 + 6 = 14$$

$$14 - 0 = 14$$

Use 🎲 🎲🎲 to show each way.

Cross out the way that does not make the number.

1. **7**	$8 - 1$	$3 + 4$	$2 + 3 + 1$
2. **15**	$7 + 6$	$15 - 0$	$8 + 7$
3. **13**	$4 + 4 + 5$	$9 - 4$	$6 + 7$
4. **9**	$8 + 2$	$3 + 3 + 3$	$10 - 1$
5. **18**	$9 + 9$	$9 - 9$	$18 - 0$

Basic Facts to 20

COMMON CORE STANDARD CC.1.OA.6
Add and subtract within 20.

Mr. Chi has 12 books.
He sells 3 books.
How many books are left?

What is 12 − 3?

THINK
I can count back.

THINK
I can use a related fact.

Start at 12.

Count 11, 10, _9_.

$3 + 9 = 12$

$12 - 3 = \underline{9}$

So, 12 − 3 = _9_.

Add or subtract.

1. $14 - 5 = \underline{}$ 2. $9 + 2 = \underline{}$ 3. $6 + 4 = \underline{}$

4. $12 - 6 = \underline{}$ 5. $8 - 3 = \underline{}$ 6. $7 + 5 = \underline{}$

7. $9 + 6 = \underline{}$ 8. $13 - 9 = \underline{}$ 9. $8 + 8 = \underline{}$

Count by Ones to 120

COMMON CORE STANDARD CC.1.NBT.1
Extend the counting sequence.

1	2	3	4	5	6	7	8	9	10
11	12	13	14	15	16	17	18	19	20
21	22	23	24	25	26	27	28	29	30
31	32	33	34	35	36	37	38	39	40
41	42	43	44	45	46	47	48	49	50
51	52	53	54	55	56	57	58	59	60
61	62	63	64	65	66	67	68	69	70
71	72	73	74	75	76	77	78	79	80
81	82	83	84	85	86	87	88	89	90
91	92	93	94	95	96	97	98	99	100
101	102	103	104	105	106	107	108	109	110
111	112	113	114	115	116	117	118	119	120

**Count forward.
Write the numbers.**

I find **108** on the chart.
109 comes next.

108, _109_, _110_, _111_

Use a Counting Chart. Count forward.
Write the numbers.

1. 112, ____, ____, ____

2. 25, ____, ____, ____

3. 95, ____, ____, ____

4. 50, ____, ____, ____

Count by Tens to 120

Use the Counting Chart.
Count forward by tens.
Start on 4.

14, 24, 34, 44, 54,

64, 74, __84__, __94__,

__104__, __114__

1	2	3	4	5	6	7	8	9	10
11	12	13	14	15	16	17	18	19	20
21	22	23	24	25	26	27	28	29	30
31	32	33	34	35	36	37	38	39	40
41	42	43	44	45	46	47	48	49	50
51	52	53	54	55	56	57	58	59	60
61	62	63	64	65	66	67	68	69	70
71	72	73	74	75	76	77	78	79	80
81	82	83	84	85	86	87	88	89	90
91	92	93	94	95	96	97	98	99	100
101	102	103	104	105	106	107	108	109	110
111	112	113	114	115	116	117	118	119	120

Use the Counting Chart to count by tens.
Write the numbers.

1. Start on 5.

15, 25, 35, 45, ____, ____, ____, ____, ____

2. Start on 38.

48, 58, 68, ____, ____, ____, ____, ____, ____

3. Start on 26.

36, 46, ____, ____, ____, ____, ____,

Understand Ten and Ones

COMMON CORE STANDARD CC.1.NBT.2b
Understand place value.

You can use ⬚ to show ten and some ones.
You can write ten and ones in different ways.

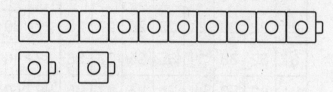

___1___ ten ___2___ ones

___10___ + ___2___

___12___

Use the model. Write the number three different ways.

1.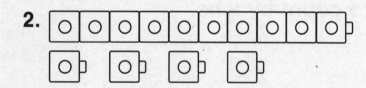

_____ ten _____ ones

_____ + _____

_____

2.

_____ ten _____ ones

_____ + _____

_____

3.

_____ ten _____ ones

_____ + _____

_____

Make Ten and Ones

COMMON CORE STANDARD CC.1.NBT.2b
Understand place value.

You can make 1 ten with 10 🎲.

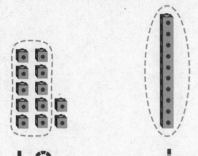

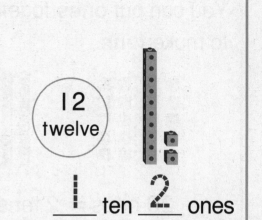

__12__ ones = __1__ ten __2__ ones

__1__ ten __2__ ones

Write how many tens and ones.

1.

15
fifteen

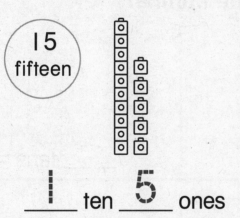

__1__ ten __5__ ones

2.

14
fourteen

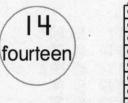

_____ ten _____ ones

3.

16
sixteen

_____ ten _____ ones

4.

13
thirteen

_____ ten _____ ones

Tens

COMMON CORE STANDARDS CC.1.NBT.2a, CC.1.NBT.2c
Understand place value.

You can put **ones** together to make **tens**.

20 ones = 2 tens

Draw to show the 2 tens.

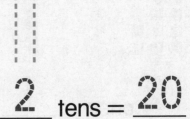

___2___ tens = 20

Use 🔳. Make groups of ten. Draw the tens. Write how many tens. Write the number.

1.

30 ones = 3 tens

_____ tens = _____

2.

40 ones = 4 tens

_____ tens = _____

3.

50 ones = 5 tens

_____ tens = _____

Tens and Ones to 50

COMMON CORE STANDARD CC.1.NBT.2
Understand place value.

You can use tens and ones
to show a number.

There are 4 tens.
There are 2 ones.
This shows 42.

Tens	Ones

4 tens   2 ones = __42__

Use your MathBoard and ⬚⬚⬚⬚⬚⬚⬚⬚ ▫ to show the tens and ones. Write the numbers.

1.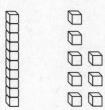

   1 tens 8 ones = __18__

2.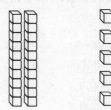

   2 tens 5 ones = _____

3.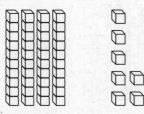

   4 tens 7 ones = _____

4.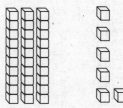

   3 tens 6 ones = _____

Tens and Ones to 100

If you know the tens and ones,
you can write the number.

Tens	Ones

There are 9 tens.
There are 8 ones.
The number is 98.

9 tens   8 ones = __98__

Use your MathBoard and ▭ ▱ **to show
the tens and ones. Write the numbers.**

1.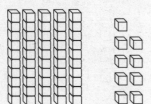

5 tens 9 ones = __59__

2.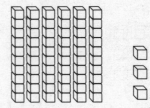

6 tens 3 ones = _____

3.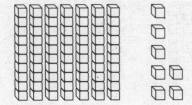

7 tens 7 ones = _____

4.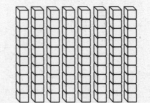

8 tens 2 ones = _____

COMMON CORE STANDARDS CC.1.NBT.2a,
CC.1.NBT.3
Understand place value.

Problem Solving • Show Numbers in Different Ways

How can you show the number 34 two different ways?

Unlock the Problem

What do I need to find?	**What information do I need to use?**
two different ways to show a number	The number is 34.

Show how to solve the problem.

> **THINK**
> You can trade I ten for I0 ones.

First Way

Tens	Ones

Second Way

Tens	Ones

1. Use ▭▭▭▭▭ ▫ to show 26 two different ways. Draw both ways.

Tens	Ones

Tens	Ones

Model, Read, and Write Numbers from 100 to 110

COMMON CORE STANDARD CC.1.NBT.1
Extend the counting sequence.

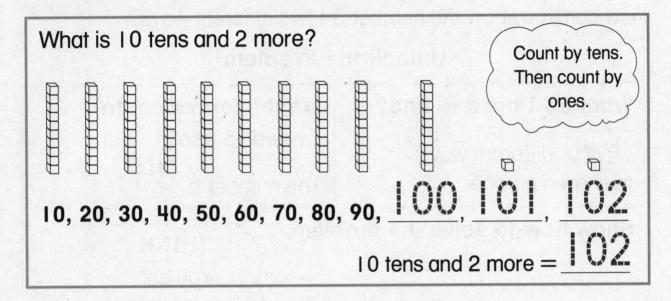

What is 10 tens and 2 more?

Count by tens. Then count by ones.

10, 20, 30, 40, 50, 60, 70, 80, 90, 100, 101, 102

10 tens and 2 more = 102

Use 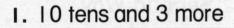 to model the number.

Write the number.

1. 10 tens and 3 more

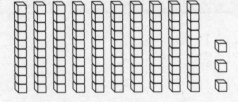

_____

2. 10 tens and 7 more

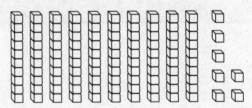

_____

3. 10 tens and 6 more

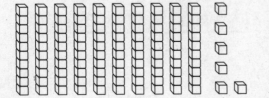

_____

4. 10 tens and 9 more

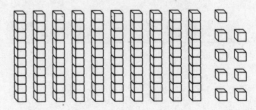

_____

Model, Read, and Write Numbers from 110 to 120

COMMON CORE STANDARD CC.1.NBT.1
Extend the counting sequence.

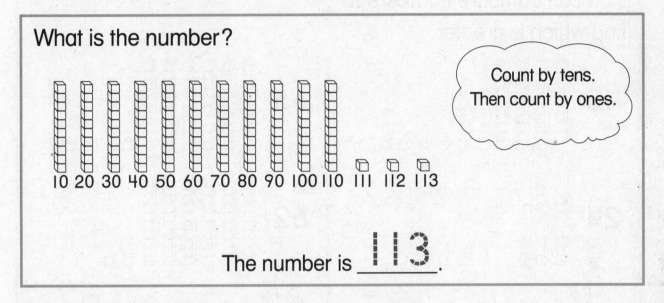

What is the number?

Count by tens.
Then count by ones.

10 20 30 40 50 60 70 80 90 100 110 111 112 113

The number is __113__.

Use to model the number.

Write the number.

I.

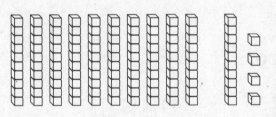

_____

2.

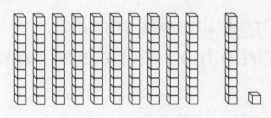

_____

3.

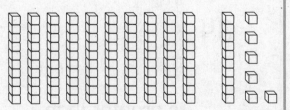

_____

4.

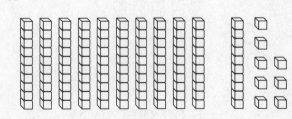

_____

Algebra • Greater Than

You can compare numbers to
find which is greater.

48

24

65

62

__48__ is greater than __24__.

__65__ is greater than __62__.

__48__ > __24__

__65__ > __62__

Draw lines to match.
Write the numbers to compare.

1.

43

55

_____ is greater than _____.

_____ > _____

2.

51

34

_____ is greater than _____.

_____ > _____

Algebra • Less Than

COMMON CORE STANDARD CC.1.NBT.3
Understand place value.

You can compare numbers to
find which is less.

23

26

23 is less than 26.

23 < 26

65

43

43 is less than 65.

43 < 65

Draw lines to match.
Write the numbers to compare.

1.

37

31

_____ is less than _____.

_____ < _____

2.

74

44

_____ is less than _____.

_____ < _____

COMMON CORE STANDARD CC.1.NBT.3
Understand place value.

Algebra • Use Symbols to Compare

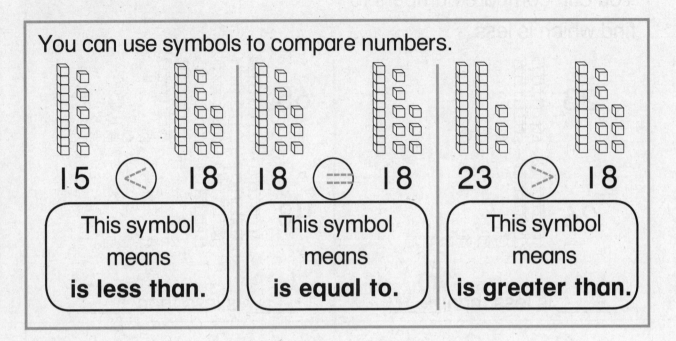

You can use symbols to compare numbers.

15 ⟨<⟩ 18

This symbol means is less than.

18 ⟨=⟩ 18

This symbol means is equal to.

23 ⟨>⟩ 18

This symbol means is greater than.

Write >, <, or =. Complete the sentence.

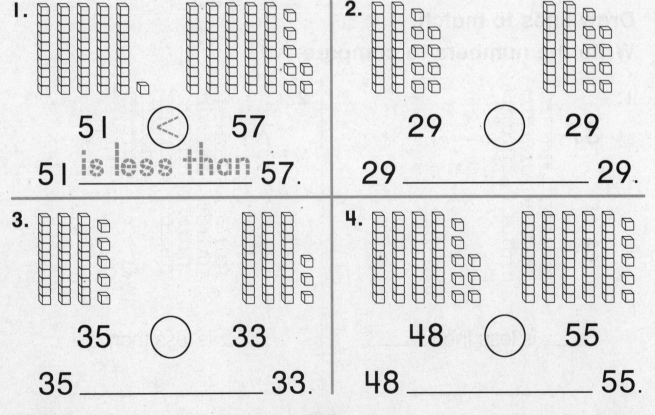

1. 51 ⟨<⟩ 57

51 is less than 57.

2. 29 ◯ 29

29 _____ 29.

3. 35 ◯ 33

35 _____ 33.

4. 48 ◯ 55

48 _____ 55.

0 more

npare

| 5 | 7 | 8 | 10 | 11 |

ds
ards
nd greater
s Anthony have now?

nlock the Problem

	What information do I need to use?
cards	number cards < __6__
s now	and > __9__.

to solve the problem.

| 8 | ~~10~~ | ~~11~~ |

> **THINK**
> Cross out the numbers
> Anthony gives away.

mber cards __7__ and __8__.

to solve.

number cards shown.
ay the cards less
greater than 22.
s does she have now?

| 20 | 21 | 23 |

Emily has _____ and _____.

ing Company

Name _____

COMMON CORE STA...
Use place value under...
properties of operatio...

10 Less, 10 More

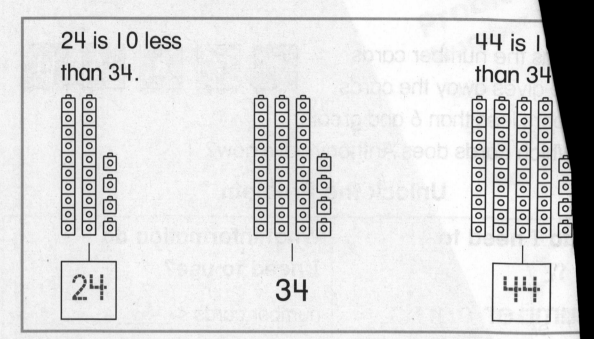

24 is 10 less
than 34.

44 is 1...
than 34...

24   34   44

Write the numbers that are
10 less and 10 more.

1.
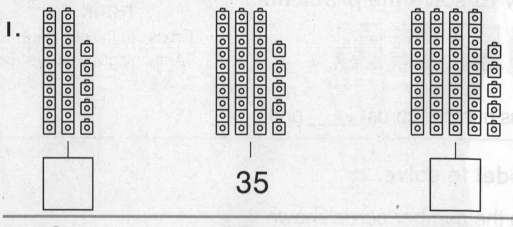

35

2.
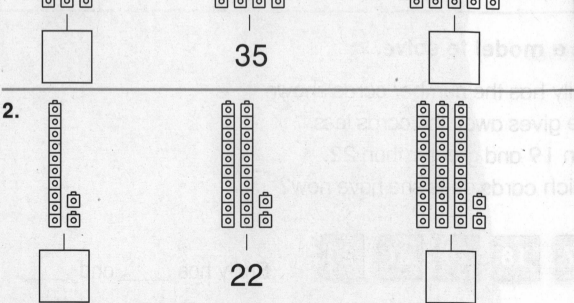

22

Add and Subtract within 20

COMMON CORE STANDARD CC.1.OA.6
Add and subtract within 20.

You can use strategies to add or subtract.

- count on
- doubles
- doubles plus one
- count back
- related facts
- doubles minus one

What is $5 + 6$?

I can use doubles plus one.

$$5 + 5 = 10$$

So, $5 + 6 = \underline{11}$.

What is $12 - 4$?

I can use a related fact.

$$8 + 4 = 12$$

So, $12 - 4 = \underline{8}$.

Add or subtract.

1. $12 - 3 =$ _____

2. $8 + 9 =$ _____

3. $10 - 5 =$ _____

4. $13 - 7 =$ _____

5. $7 + 8 =$ _____

6. $6 + 6 =$ _____

Add Tens

COMMON CORE STANDARD CC.1.NBT.4
Use place value understanding and
properties of operations to add and subtract.

What is 10 + 30?

| | | |

Use ⬡⬡⬡⬡⬡.
Start with 1 ten.
Add 3 more tens.
Draw the tens.

1 ten + 3 tens = __4__ tens

10 + 30 = __40__

**Use ⬡⬡⬡⬡⬡. Draw to show tens.
Write how many tens. Write the sum.**

1.

1 ten + 8 tens = _____ tens

10 + 80 = ____

2.

4 tens + 3 tens = _____ tens

40 + 30 = ____

3.

2 tens + 6 tens = _____ tens

20 + 60 = ____

4.

5 tens + 3 tens = _____ tens

50 + 30 = ____

Subtract Tens

COMMON CORE STANDARD CC.1.NBT.6
Use place value understanding and
properties of operations to add and subtract.

What is 60 − 40?

6 tens − 4 tens = __2__ tens

60 − 40 = __20__

Use .
Show 6 tens.
Take away 4 tens.
2 tens are left.

Use ▭▭▭▭. **Draw to show tens.**
Write how many tens. Write the difference.

1.

7 tens − 4 tens = _____ tens

70 − 40 = ___

2.

9 tens − 5 tens = _____ tens

90 − 50 = ___

3.

5 tens − 2 tens = _____ tens

50 − 20 = ___

4.

8 tens − 7 tens = _____ ten

80 − 70 = ___

COMMON CORE STANDARD CC.1.NBT.4

Use a Hundred Chart to Add

Use place value understanding and properties of operations to add and subtract.

You can count on to add on a hundred chart.

1	2	3	4	5	6	7	8	9	10
11	12	13	14	15	16	17	18	19	20
21	22	23	24	25	26	27	28	29	30
31	32	33	34	35	36	37	38	39	40
41	42	43	44	45	46	47	48	49	50
51	52	53	54	55	56	57	58	59	60
61	62	63	64	65	66	67	68	69	70
71	72	73	74	75	76	77	78	79	80
81	82	83	84	85	86	87	88	89	90
91	92	93	94	95	96	97	98	99	100

Start at 21. Move right to count on 3 ones. Count

22 , 23 , 24

$21 + 3 =$ 24

Start at 68. Move down to count on 3 tens. Count

78 , 88 , 98

$68 + 30 =$ 98

Use the hundred chart to add.

Count on by ones.

1. $46 + 2 =$ _____

2. $63 + 3 =$ _____

Count on by tens.

3. $52 + 30 =$ _____

4. $23 + 40 =$ _____

Use Models to Add

COMMON CORE STANDARD CC.1.NBT.4
Use place value understanding and
properties of operations to add and subtract.

Add ones to a two-digit number.

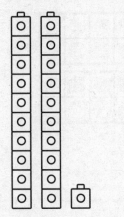

THINK
Draw 2 tens
and 4 ones.

$$21 \quad + \quad 3 \quad = \quad \underline{24}$$

Add tens to a two-digit number.

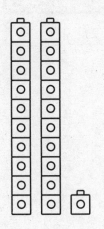

THINK

$$21 \quad + \quad 30 \quad = \quad \underline{51}$$

**Use ⬜. Draw to show how to add the
ones or tens. Write the sum.**

1. $15 + 2 = $ _____

2. $15 + 20 = $ _____

Make Ten to Add

COMMON CORE STANDARD CC.1.NBT.4
Use place value understanding and
properties of operations to add and subtract.

What is 17 + 5?

Step 1

Use ⬤.

Show 17.

Use ◯.

Show 5.

Step 2

Make
a
ten.

Step 3 Add.

$$20 + 2 = \underline{22}$$

$$\text{So, } 17 + 5 = \underline{22}.$$

Draw to show how you make a ten. Find the sum.

1. What is 16 + 8?

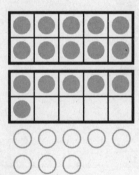

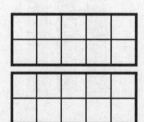

$$\underline{20} + \underline{4} = \underline{24}$$

$$\text{So, } 16 + 8 = \underline{}.$$

Use Place Value to Add

COMMON CORE STANDARD CC.1.NBT.4
Use place value understanding and
properties of operations to add and subtract.

You can use tens and ones to help you add.

Add 25 and 22.

Show 25. ⟶

Show 22. ⟶

Tens	Ones

How many tens? 2 tens + 2 tens = __4__ tens

How many ones? 5 ones + 2 ones = __7__ ones

__4__ tens + __7__ ones

$$40 + 7 = 47$$

$$\begin{array}{r} 25 \\ +22 \\ \hline 47 \end{array}$$

Use tens and ones to add.

1. Add 34 and 42.

Tens	Ones

3 tens + 4 tens = _____ tens

4 ones + 2 ones = _____ ones

_____ tens + _____ ones

_____ + _____ = _____

$$\begin{array}{r} 34 \\ +42 \\ \hline \end{array}$$

COMMON CORE STANDARD CC.1.NBT.4
Use place value understanding and
properties of operations to add and subtract.

Problem Solving • Addition Word Problems

Morgan plants 17 seeds.

Amy plants 8 seeds.

How many seeds do they plant?

Unlock the Problem

What do I need to find?	**What information do I need to use?**
how many _seeds_ they plant	Morgan plants __17__ seeds. Amy plants __8__ seeds.

Show how to solve the problem.

count on ones

(make a ten)

add tens and ones

_____ seeds.

Draw to solve. Circle your reasoning.

1. Edward buys 24 tomato plants.

   He buys 15 pepper plants.

   How many plants does he buy?

   count on tens

   make a ten

   add tens and ones

   _____ plants

Practice Addition and Subtraction

COMMON CORE STANDARDS CC.1.NBT.4, CC.1.NBT.6
Use place value understanding and properties of
operations to add and subtract.

You can use models to add and subtract.

$$13 + 5 = \underline{18}$$

$$90 - 60 = \underline{30}$$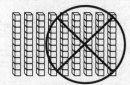

Add or subtract.

1. $33 + 6 =$ ___	2. $10 + 10 =$ ___	3. $15 - 8 =$ ___
4. $6 + 7 =$ ___	5. $54 + 23 =$ ___	6. $71 + 8 =$ ___
7. $5 + 5 =$ ___	8. $8 - 8 =$ ___	9. $16 + 3 =$ ___
10. $55 + 12 =$ ___	11. $9 - 7 =$ ___	12. $30 - 10 =$ ___

Order Length

COMMON CORE STANDARD CC.1.MD.1
Measure lengths indirectly and by iterating
length units.

You can put objects in order by length.

These pencils are in order from **shortest** to **longest**.	These pencils are in order from **longest** to **shortest**.
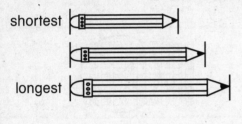	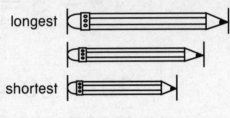

Draw three lines in order from **shortest** to **longest**.

1. shortest |

2. |

3. longest |

Draw three lines in order from **longest** to **shortest**.

4. longest |

5. |

6. shortest |

me _____

COMMON CORE STANDARD CC.1.MD.2
Measure lengths indirectly and by iterating
length units.

se Nonstandard Units
Measure Length

ou can use  to measure length.

ine up the ▪.

ount how many.

bout __5__ ▪

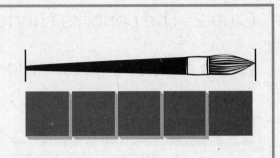

e real objects. Use ▪ to measure.

unt how many.

about _____ ▪

about _____ ▪

about _____ ▪

about _____ ▪

Indirect Measurement

COMMON CORE STANDARD
Measure lengths indirectly ar
length units.

Clue 1: A marker is shorter than a pencil.

Clue 2: The pencil is shorter than a ribbon.

Is the marker shorter or longer than the ribbon?

marker

pencil

ribbon

Draw Clu
Draw Clu
Then compo
marker and the

So, the marker is _shorter_ than the ribbon.

Use the clues. Write **shorter** or **longer**
to complete the sentence. Then draw to
prove your answer.

Draw Clu
Draw Clu
Then compo
string and the

1. Clue 1: A string is longer than a straw.

   Clue 2: The straw is longer than a pencil.

   Is the string shorter or longer than the pencil?

string

straw

pencil

The string is _____ than the pencil.

Make a Nonstandard Measuring Tool

COMMON CORE STANDARD CC.1.MD.2
Measure lengths indirectly and by iterating length units.

About how long is the ribbon?
Count to measure.

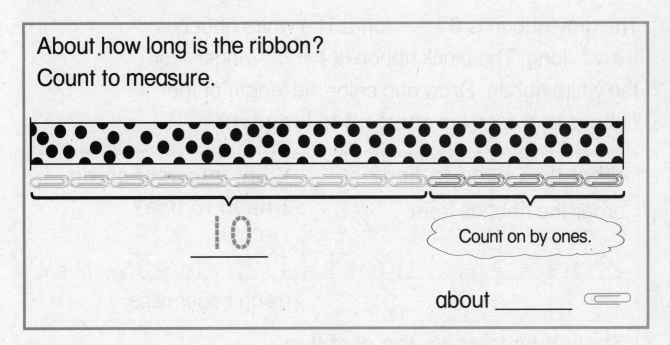

__10__

Count on by ones.

about _____

Use real objects and the measuring tool you made. Measure.

1.

about _____ ⌒

2.

about _____ ⌒

3.

about _____ ⌒

Problem Solving •
Measure and Compare

COMMON CORE STANDARD CC.1.MD.2
Measure lengths indirectly and by iterating length units.

The gray ribbon is 3 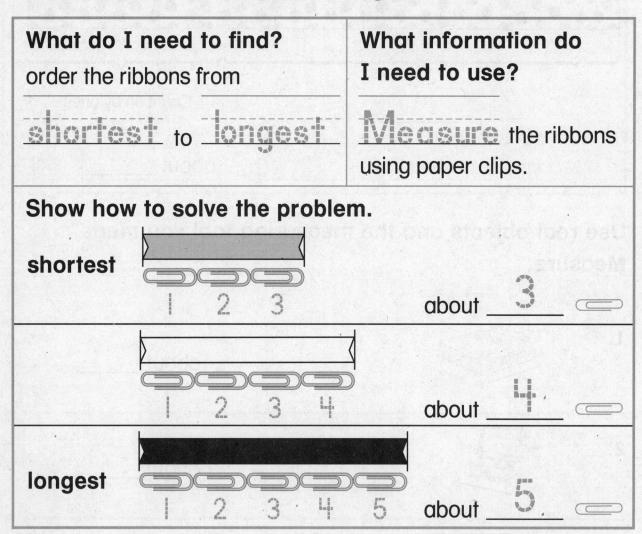 long. The white ribbon is 4 ⌒ long. The black ribbon is 1 ⌒ longer than the white ribbon. Draw and color the length of the ribbons in order from **shortest** to **longest**.

What do I need to find?	**What information do I need to use?**
order the ribbons from	
shortest to _longest_	_Measure_ the ribbons using paper clips.

Show how to solve the problem.

shortest

about __3__ ⌒

about __4__ ⌒

longest

about __5__ ⌒

1. The _____ ribbon is the shortest ribbon.

2. The _____ ribbon is the longest ribbon.

Time to the Hour

Look at the hour hand.

The hour hand points to the __**8**__.

It is __8:00__.

Look at where the hour hand points.
Write the time.

1. The hour hand points to the _____.

   It is _____.

2. The hour hand points to the _____.

   It is _____.

3.	4.	5.

_____	_____	_____

Name _____

Time to the Half Hour

COMMON CORE STANDARD CC.1.MD.3
Tell and write time.

The hour hand points halfway between

the __9__ and the __10__.

It is __half past 9:00__.

Look at where the hour hand points.
Write the time.

1. The hour hand points halfway between

   the _____ and the _____.

   It is _____.

---

2. The hour hand points halfway between

   the _____ and the _____.

   It is _____.

3.

_____

4.

_____

5.

_____

Tell Time to the Hour and Half Hour

COMMON CORE STANDARD CC.1.MD.3
Tell and write time.

The short hand is the **hour hand**.
It shows the hour.

The long hand is the **minute hand**.
It shows the minutes after the hour.

There are 60 minutes
in one hour.

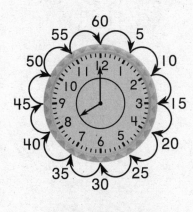

There are 30 minutes
in a half hour.

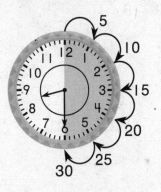

Write the time.

1.

2.

3.

Practice Time to the Hour and Half Hour

COMMON CORE STANDARD CC.1.MD.3
Tell and write time.

The hour hand points to 8.

The minute hand points to 12.

8:00

The hour hand points between 8 and 9.

The minute hand points to 6.

8:30

Use the hour hand to write the time.
Draw the minute hand.

1.

_____

2.

_____

3.

_____

COMMON CORE STANDARD CC.1.MD.4
Represent and interpret data.

Read Picture Graphs

A **picture graph** uses pictures to show how many.

Count the 🯅 in each row.

Snack We Like					
🍎 apple	🯅	🯅	🯅	🯅	🯅
🥨 pretzel	🯅	🯅	🯅		

Each 🯅 stands for 1 child who chose that snack.

There are ___5___ children who chose 🍎 .

There are ___3___ children who chose 🥨 .

Use the picture graph to answer each question.

What We Ate for Lunch							
🥪 sandwich	🯅	🯅	🯅	🯅	🯅	🯅	
🥫 soup	🯅	🯅					

Each 🯅 stands for 1 child.

1. Which lunch did more
   children choose? Circle.

2. How many children chose ? _____ children

3. How many children chose ? _____ children

Make Picture Graphs

COMMON CORE STANDARD CC.1.MD.4
Represent and interpret data.

Are there more black cars or white cars?
Complete the picture graph to find out.

Cross out each car as you count.

Draw a ◯ in the graph to show each car.

Black and White Cars											
🚗 black	◌										
🚗 white											

Each ◯ stands for 1 car.

Use the picture graph to answer each question.

1. How many are there? _____

2. How many 🚗 are there? _____

3. Are there more 🚗 or ? Circle.

Name ______________________________

COMMON CORE STANDARD CC.1.MD.4
Represent and interpret data.

Read Bar Graphs

A **bar graph** uses a bar to show how many.

This graph shows 6 children chose 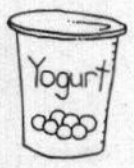 .

The longest bar shows the snack most children chose.

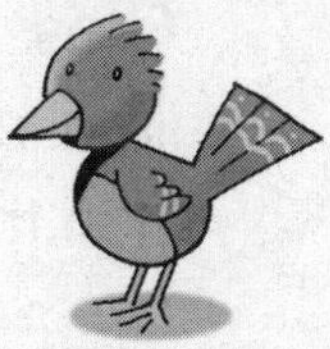

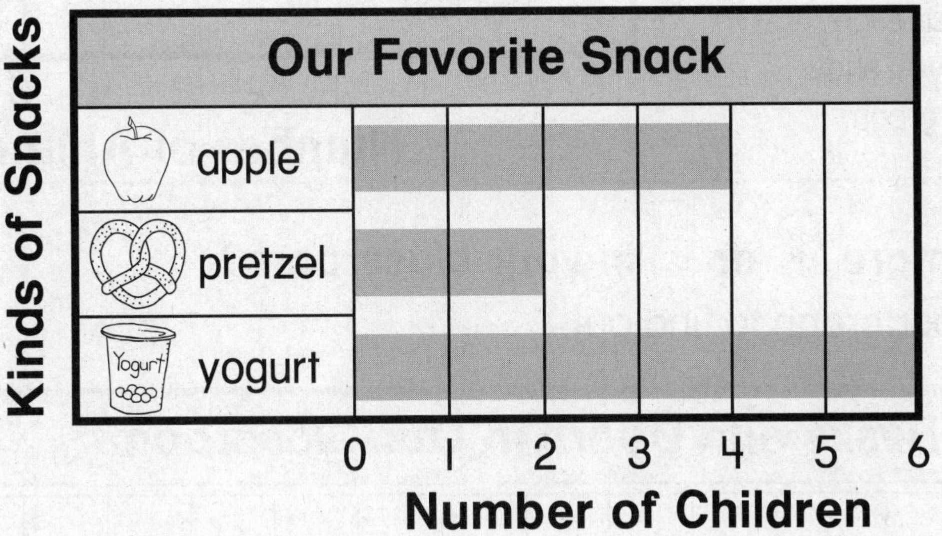
Our Favorite Snack
Kinds of Snacks
apple
pretzel
yogurt
0 1 2 3 4 5 6
Number of Children

Use the bar graph to answer the question.

1. How many children chose 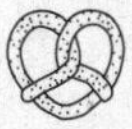? ________ children

2. How many children chose 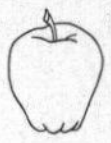? ________ children

3. Circle the snack the most children chose.

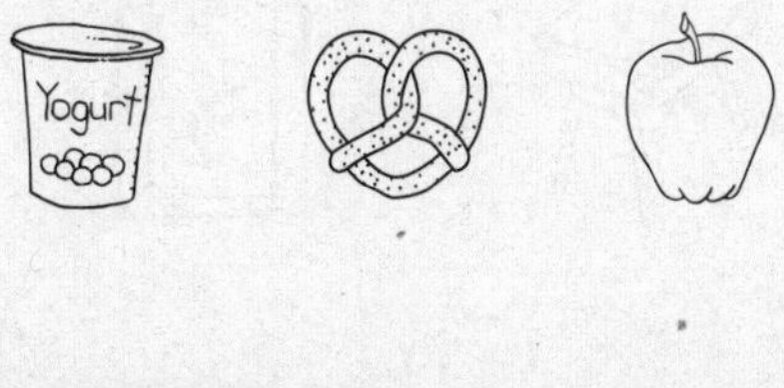

4. Circle the snack the fewest children chose.

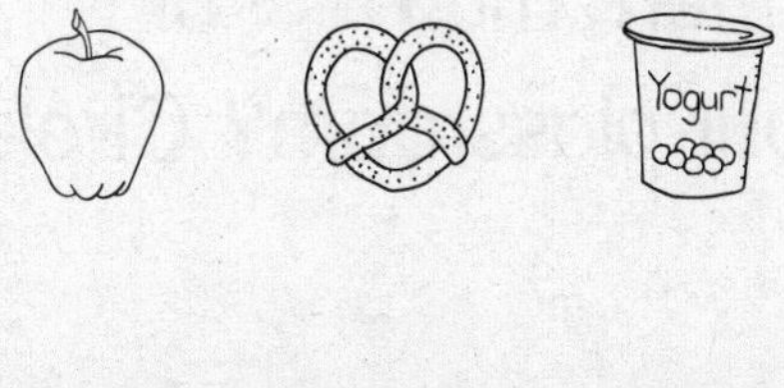

Make Bar Graphs

COMMON CORE STANDARD CC.1.MD.4
Represent and interpret data.

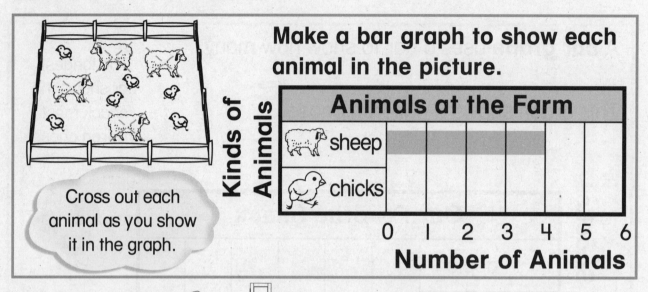

Cross out each animal as you show it in the graph.

Make a bar graph to show each animal in the picture.

Kinds of Animals

Animals at the Farm						
🐑 sheep						
🐥 chicks						

0   1   2   3   4   5   6
Number of Animals

Are there more  **or** ▯ **in your classroom?**

1. Make a bar graph to find out.

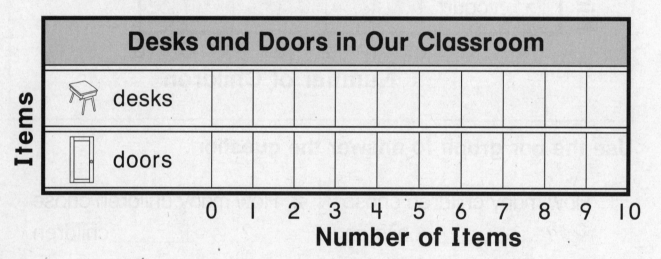

Items

Desks and Doors in Our Classroom										
🪑 desks										
▯ doors										

0   1   2   3   4   5   6   7   8   9   10
Number of Items

2. How many are in your classroom? _____

3. Are there more or ▯
   in your classroom? Circle.

COMMON CORE STANDARD CC.1.MD.4
Represent and interpret data.

Read Tally Charts

Some children named their favorite collections.

Each | stands for 1 child.

Each ‖‖‖ stands for 5 children.

Our Favorite Thing to Collect		Total
🐚 shells	|||| 1 2 3 4	4
🚗 stamps	‖‖‖ || 5 6 7	7

More children like to collect __stamps__.

Complete the tally chart.

Do you have a pet?		Total
yes	‖‖‖ |||	
no	‖‖‖	

Use the tally chart to answer each question.

1. How many children have a pet? _____ children

2. How many children do not have a pet? _____ children

3. Did more children answer yes or no? _____

Make Tally Charts

The picture shows shapes.
Make a tally chart to show
how many of each shape.

Cross out each
shape as you
count.

Shapes in the Picture		Total
◯ circles	‖‖ ‖	6
☆ stars	‖‖	3
△ triangles	‖‖ ‖‖	8

Use the tally chart to answer each question.

1. How many ☆ are there?

_____ ☆

2. How many more △ than ◯ are there?

_____ more △

3. Which shape is there the most of? Circle.

COMMON CORE STANDARD CC.1.MD.4
Represent and interpret data.

Problem Solving •
Represent Data

Ava has these beads to make a bracelet.
How can you find how many beads she has?

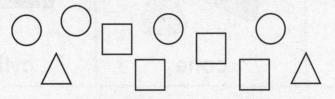

Unlock the Problem

What do I need to find?	**What information do I need to use?**
how many  Ava has	the number of ,_____, _____ and _____ in the picture

Show how to solve the problem.

Color the first bar to show there are 4 circles.

Beads Ava Has

circle ○						
square □						
triangle △						

0 1 2 3 4 5 6

Use the graph. Write how many. Add to solve.

1. ___4___ ○ + _____ □ + _____ △ = _____

How many beads does Ava have? _____ beads

Three-Dimensional Shapes

COMMON CORE STANDARD CC.1.G.1
Reason with shapes and their attributes.

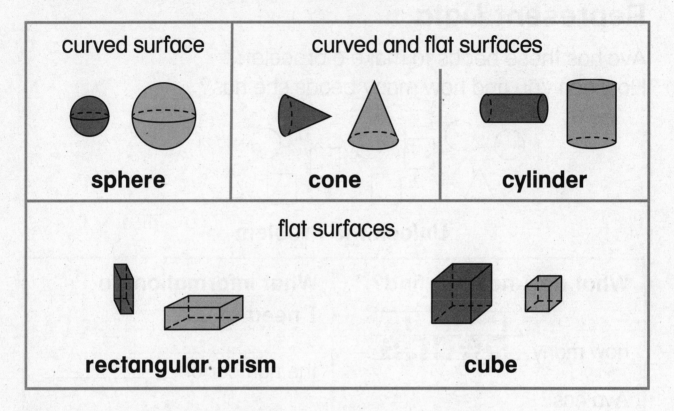

curved surface

sphere

curved and flat surfaces

cone

cylinder

flat surfaces

rectangular prism

cube

Color to sort the shapes into three groups.

1. only **flat surfaces**  RED

2. only a **curved surface** BLUE

3. both **curved** and **flat surfaces** YELLOW

cone

cube

cylinder

sphere

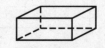

rectangular prism

COMMON CORE STANDARD CC.1.G.2
Reason with shapes and their attributes.

Combine
Three-Dimensional Shapes

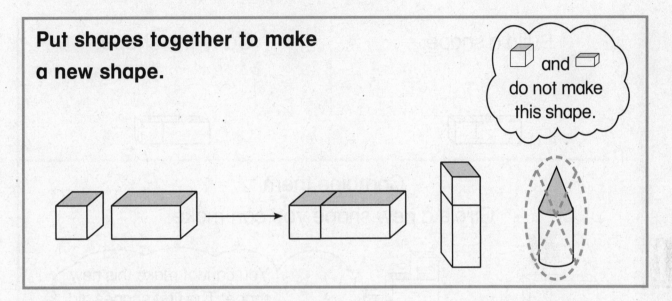

**Put shapes together to make
a new shape.**

and
do not make
this shape.

Use three-dimensional shapes.

Combine.	Which new shapes can you make? Circle them.
I.	
2.	

Make New
Three-Dimensional Shapes

COMMON CORE STANDARD CC.1.G.2
Reason with shapes and their attributes.

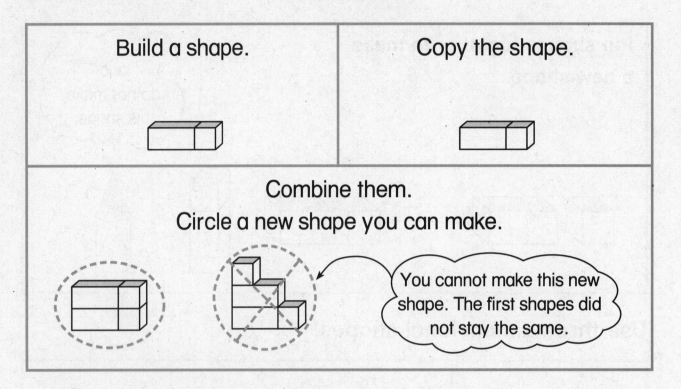

Build a shape.

Copy the shape.

Combine them.
Circle a new shape you can make.

You cannot make this new shape. The first shapes did not stay the same.

Use three-dimensional shapes.

Build these shapes.	Circle the new shape you can make. Cross out the shape you cannot make.	
I.		
2.		

COMMON CORE STANDARD CC.1.G.2
Reason with shapes and their attributes.

Problem Solving • Take Apart Three-Dimensional Shapes

Kate has △, ⬡, ▯, and ▯.
She built a tower.
Which shapes did Kate
use to build the tower?

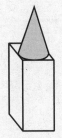

Unlock the Problem

What do I need to find?	**What information do I need to use?**
which **shapes** Kate used to build the tower	Kate has these shapes.

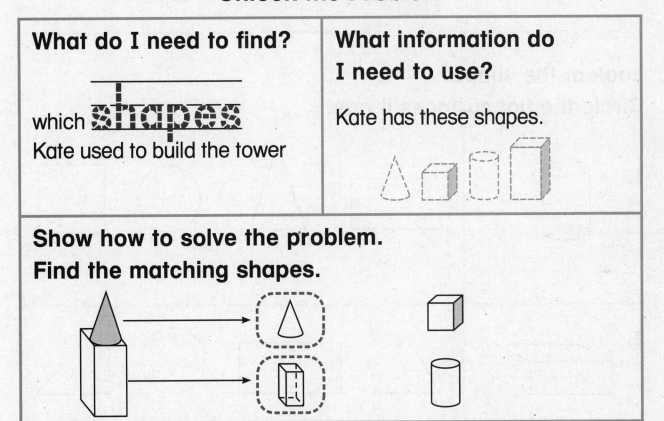

Show how to solve the problem.
Find the matching shapes.

Use three-dimensional shapes. Circle your answer.

1. Which shapes did Marvin use to build this bench?

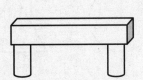

Two-Dimensional Shapes
on Three-Dimensional Shapes

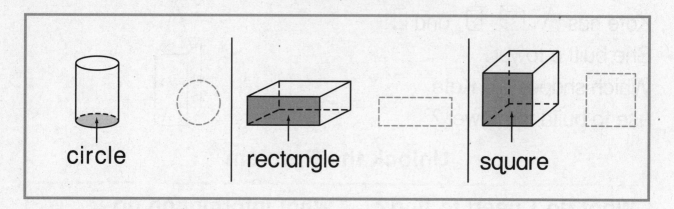

circle   rectangle   square

Look at the shape.
Circle the flat surfaces it has.

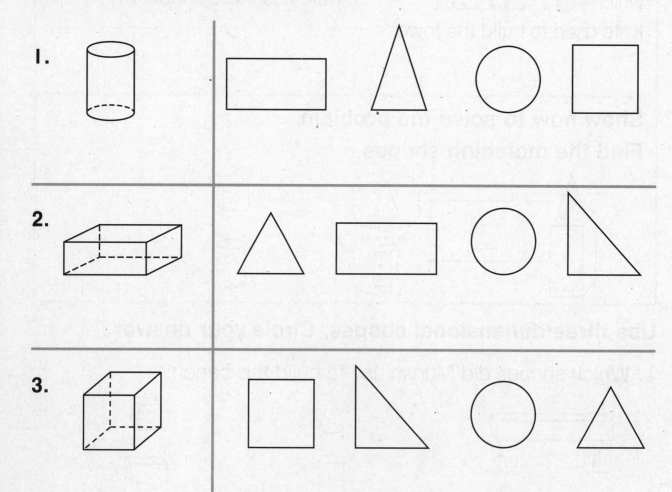

1.

2.

3.

Sort Two-Dimensional Shapes

COMMON CORE STANDARD CC.1.G.1
Reason with shapes and their attributes.

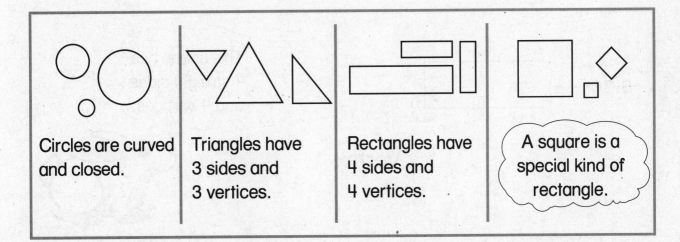

| Circles are curved and closed. | Triangles have 3 sides and 3 vertices. | Rectangles have 4 sides and 4 vertices. | A square is a special kind of rectangle. |

Read the sorting rule. Circle the shapes that follow the rule.

1. 4 sides

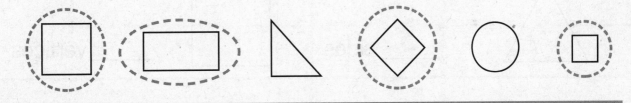

2. curved and closed

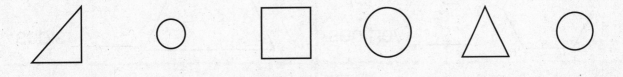

3. 3 vertices

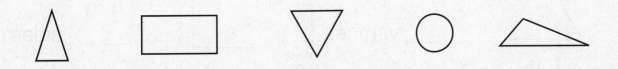

Describe Two-Dimensional Shapes

COMMON CORE STANDARD CC.1.G.1
Reason with shapes and their attributes.

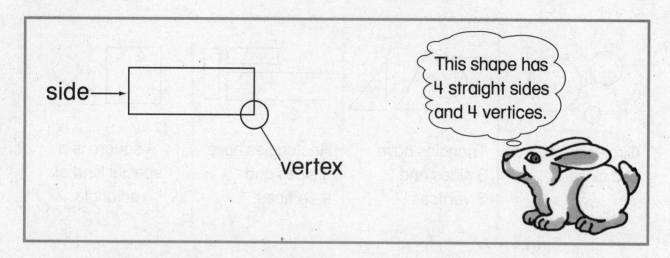

side →

vertex

This shape has 4 straight sides and 4 vertices.

Write the number of straight sides or vertices.

I. triangle

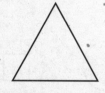

___3___ sides

2. square

_____ vertices

3. hexagon

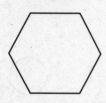

_____ vertices

4. trapezoid

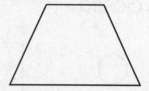

_____ sides

5. triangle

_____ vertices

6. square

_____ sides

Combine Two-Dimensional Shapes

COMMON CORE STANDARD CC.1.G.2
Reason with shapes and their attributes.

You can put shapes together to make
a new shape.

3 ____ △ make
a ▽ .

Use pattern blocks. Draw to show the blocks.
Write how many blocks you used.

1. How many ▽ make a ⬡ ?

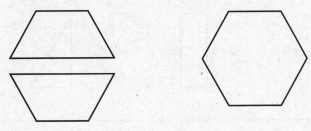

____ ▽ make a ⬡ .

2. How many ◇ make a ⬡ ?

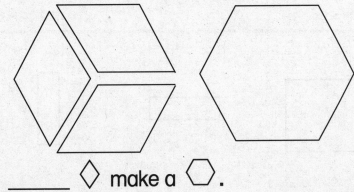

____ ◇ make a ⬡ .

Combine More Shapes

COMMON CORE STANDARD CC.1.G.2
Reason with shapes and their attributes.

Combine shapes to make a new shape.

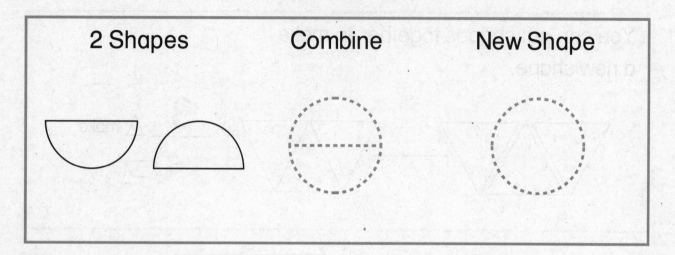

2 Shapes	Combine	New Shape

Circle the shapes that can combine to make the new shape.

1.

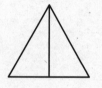

2.

3.

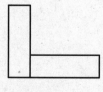

COMMON CORE STANDARD CC.1.G.2
Reason with shapes and their attributes.

Problem Solving • Make New Two-Dimensional Shapes

Luis wants to use △ to make a ◇.
How many △ does he need?

Unlock the Problem

What do I need to find?	What information do I need to use?
how Luis can make a _____ using _____	Luis uses

Show how to solve the problem.

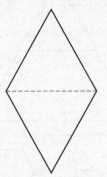

_2_ △ make a ◇.

Use shapes to solve.

1. Meg wants to use △
   to make a ▱.
   _____ △ make a ▱.

Find Shapes in Shapes

COMMON CORE STANDARD CC.1.G.2
Reason with shapes and their attributes.

Which two pattern blocks make this shape?

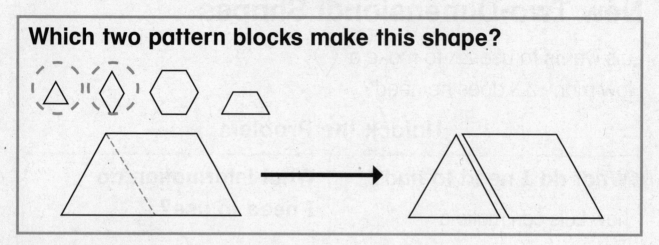

Use two pattern blocks to make the shape.
Circle the blocks you use.

1.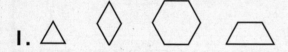

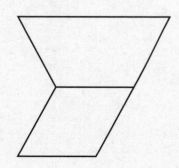

2.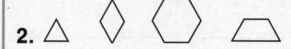

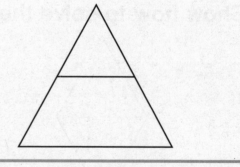

3.

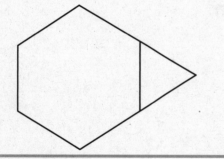

4.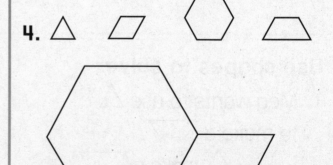

Take Apart
Two-Dimensional Shapes

COMMON CORE STANDARD CC.1.G.2
Reason with shapes and their attributes.

Use pattern blocks to help you find the parts of a shape.

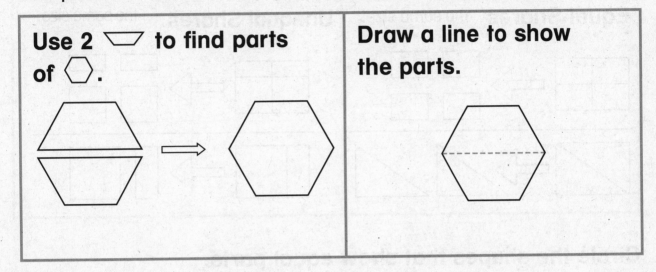

Use 2 ⬭ to find parts of ⬡.

Draw a line to show the parts.

Use pattern blocks. Draw a line to show the parts.

1. Show 2 △.

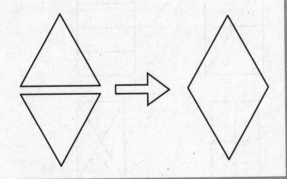

2. Show 2 ⬭.

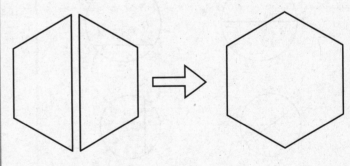

3. Show 2 ☐.

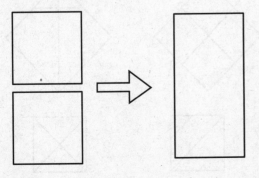

4. Show 2 ◠.

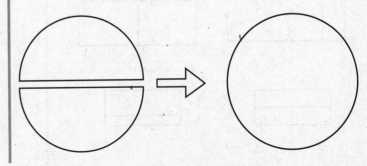

Equal or Unequal Parts

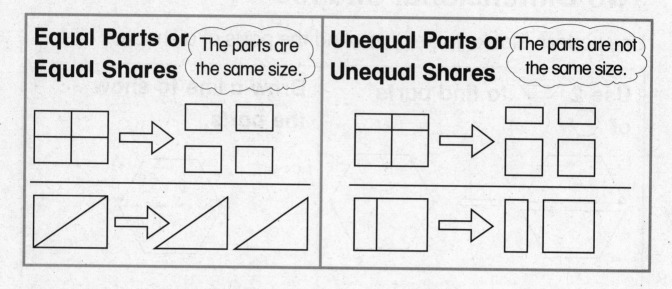

| Equal Parts or Equal Shares | The parts are the same size. |
| Unequal Parts or Unequal Shares | The parts are not the same size. |

Circle the shapes that show equal parts.
Cross out the shapes that show unequal parts.

1.

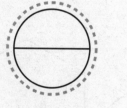

2.

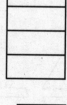

3.

4.

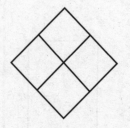
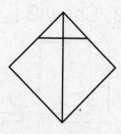

Name _____

Halves

COMMON CORE STANDARD CC.1.G.3
Reason with shapes and their attributes.

How can you show **halves**?

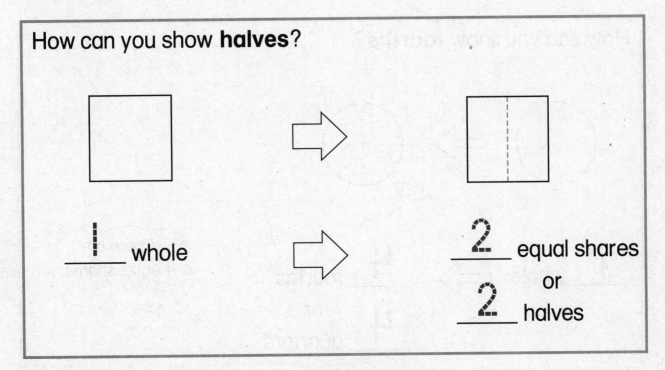

_____ whole

_____ equal shares
or
_____ halves

Draw a line to show halves. Write the numbers.

I.

_____ whole   _____ halves

2.

_____ whole   _____ halves

Fourths

COMMON CORE STANDARD CC.1.G.3
Reason with shapes and their attributes.

How can you show fourths?

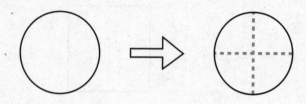

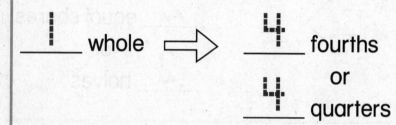

_____ whole ⟹ 4 _____ fourths
or
4 _____ quarters

There are
4 equal shares.

Draw lines to show fourths. Write the number.

1.

_____ whole   _____ fourths

2. Draw lines to show quarters. Write the number.

_____ whole   _____ fourths